ENERGY SECTOR STANDARD
OF THE PEOPLE'S REPUBLIC OF CHINA
中华人民共和国能源行业标准

Archives Acceptance Specification for Wind Power Projects

风电场工程档案验收规程

NB/T 31118-2017

Chief Development Department: China Renewable Energy Engineering Institute

Approval Department: National Energy Administration of the People's Republic of China

Implementation Date: March 1, 2018

China Water & Power Press

中国水利水电出版社

Beijing 2024

All rights reserved. No part of this publication may be reproduced, stored in a retrieval system, or transmitted in any form or by any means—electronic, mechanical, photocopying, recording or otherwise, without prior written permission of the publisher.

图书在版编目（CIP）数据

风电场工程档案验收规程 = Archives Acceptance Specification for Wind Power Projects (NB/T 31118-2017)：英文 / 国家能源局发布. -- 北京：中国水利水电出版社, 2024. 5. -- ISBN 978-7-5226-2590-4

Ⅰ. TM614-65

中国国家版本馆CIP数据核字第2024LZ7704号

ENERGY SECTOR STANDARD
OF THE PEOPLE'S REPUBLIC OF CHINA
中华人民共和国能源行业标准

Archives Acceptance Specification
for Wind Power Projects
风电场工程档案验收规程
NB/T 31118-2017
（英文版）

Issued by National Energy Administration of the People's Republic of China
国家能源局　发布
Translation organized by China Renewable Energy Engineering Institute
水电水利规划设计总院　组织翻译
Published by China Water & Power Press
中国水利水电出版社　出版发行
　　Tel: (+ 86 10) 68545888　68545874
　　sales@mwr.gov.cn
　　Account name: China Water & Power Press
　　Address: No.1, Yuyuantan Nanlu, Haidian District, Beijing 100038, China
　　http://www.waterpub.com.cn
中国水利水电出版社微机排版中心　排版
北京中献拓方科技发展有限公司　印刷
184mm×260mm　16开本　3印张　95千字
2024年5月第1版　2024年5月第1次印刷
Price（定价）：￥480.00

Introduction

This English version is one of China's energy sector standard series in English. Its translation was organized by China Renewable Energy Engineering Institute authorized by National Energy Administration of the People's Republic of China in compliance with relevant procedures and stipulations. This English version was issued by National Energy Administration of the People's Republic of China in Announcement [2021] No. 5 dated November 16, 2021.

This version was translated from the Chinese Standard NB/T 31118-2017, *Archives Acceptance Specification for Wind Power Projects*, published by China Water & Power Press. The copyright is reserved by National Energy Administration of the People's Republic of China. In the event of any discrepancy in the implementation, the Chinese version shall prevail.

Many thanks go to the staff from relevant standard development organizations and those who have provided generous assistance in the translation and review process.

For further improvement of the English version, any comments and suggestions are welcome and should be addressed to:

China Renewable Energy Engineering Institute
No. 2 Beixiaojie, Liupukang, Xicheng District, Beijing 100120, China
Website: www.creei.cn

Translating organization:

POWERCHINA Zhongnan Engineering Corporation Limited

Translating staff:

LI Qian   CHEN Lei   YANG Hong

Review panel members:

QIE Chunsheng	Senior English Translator
JIN Feng	Tsinghua University
YAN Wenjun	Army Academy of Armored Forces, PLA
QI Wen	POWERCHINA Beijing Engineering Corporation Limited
QIAO Peng	POWERCHINA Northwest Engineering Corporation Limited
LIU Xiaofen	POWERCHINA Zhongnan Engineering Corporation

	Limited
MI Biao	Yalong River Hydropower Development Co., Ltd.
LI Yu	POWERCHINA Huadong Engineering Corporation Limited
WANG Lingling	POWERCHINA Zhongnan Engineering Corporation Limited

National Energy Administration of the People's Republic of China

翻译出版说明

本译本为国家能源局委托水电水利规划设计总院按照有关程序和规定，统一组织翻译的能源行业标准英文版系列译本之一。2021年11月16日，国家能源局以2021年第5号公告予以公布。

本译本是根据中国水利水电出版社出版的《风电场工程档案验收规程》NB/T 31118—2017翻译的，著作权归国家能源局所有。在使用过程中，如出现异议，以中文版为准。

本译本在翻译和审核过程中，本标准编制单位及编制组有关成员给予了积极协助。

为不断提高本译本的质量，欢迎使用者提出意见和建议，并反馈给水电水利规划设计总院。

  地址：北京市西城区六铺炕北小街2号
  邮编：100120
  网址：www.creei.cn

本译本翻译单位：中国电建集团中南勘测设计研究院有限公司

本译本翻译人员：李 倩 陈 蕾 杨 虹

本译本审核人员：

  郄春生 英语高级翻译

  金 峰 清华大学

  闫文军 中国人民解放军陆军装甲兵学院

  齐 文 中国电建集团北京勘测设计研究院有限公司

  乔 鹏 中国电建集团西北勘测设计研究院有限公司

  刘小芬 中国电建集团中南勘测设计研究院有限公司

  米 彪 雅砻江流域水电开发有限公司

  李 瑜 中国电建集团华东勘测设计研究院有限公司

  王玲玲 中国电建集团中南勘测设计研究院有限公司

国家能源局

Announcement of National Energy Administration of the People's Republic of China [2017] No. 10

According to the requirements of Document GNJKJ [2009] No. 52, "Notice on Releasing the Energy Sector Standardization Administration Regulations (*tentative*) and detailed implementation rules issued by National Energy Administration of the People's Republic of China", 204 sector standards such as *Safety Management Specification of Coalbed Methane Production Station*, including 62 energy standards (NB), 86 electric power standards (DL), and 56 petroleum standards (SY), are issued by National Energy Administration of the People's Republic of China after due review and approval.

Attachment: Directory of Sector Standards

National Energy Administration of the People's Republic of China

November 15, 2017

Attachment:

Directory of Sector Standards

Serial number	Standard No.	Title	Replaced standard No.	Adopted international standard No.	Approval date	Implementation date
...						
15	NB/T 31118-2017	Archives Acceptance Specification for Wind Power Projects			2017-11-15	2018-03-01
...						

Foreword

According to the requirements of Document GNKJ [2014] No. 298 issued by National Energy Administration of the People's Republic of China, "Notice on Releasing the Development and Revision Plan of the First Batch of Energy Sector Standards in 2014", and after extensive investigation and research, summarization of practical experience, and wide solicitation of opinions, the drafting group has prepared this specification.

The main technical contents of this specification include: acceptance organization, acceptance preconditions, and acceptance procedures.

National Energy Administration of the People's Republic of China is in charge of the administration of this specification. China Renewable Energy Engineering Institute has proposed this specification and is responsible for its routine management. Sub-committee on Construction and Installation of Wind Power Project of Energy Sector Standardization Technical Committee on Wind Power is responsible for the interpretation of the specific technical content. Comments and suggestions in the implementation of this specification should be addressed to:

China Renewable Energy Engineering Institute
No. 2 Beixiaojie, Liupukang, Xicheng District, Beijing 100120, China

Chief development organizations:

China Renewable Energy Engineering Institute

HydroChina Corporation

POWERCHINA Zhongnan Engineering Corporation Limited

Participating development organizations:

China Datang Corporation Shandong Yantai Electric Power Development Co., Ltd.

Huanghe Hydropower Development Co., Ltd.

Chief drafting staff:

WU Hehe	FAN Jianzhen	NI Ping	ZHANG Dingrong
LI Tuqiang	FU Zhengning	ZHANG Haibin	DENG Xuehui
LENG Hui	CHEN Zhongxing		

Review panel members:

YI Yuechun	WANG Yanmin	WANG Hongmin	CHANG Zuowei

XIE Hongwen	HOU Hongying	LIU Zhifang	QIN Chusheng
LIU Wei	ZHANG Qingyuan	WANG Liqun	YU Hui
YAO Hui	LI Jian	LI Yan	LI Shisheng

Contents

1	General Provisions	1
2	Terms	2
3	Acceptance Organization	3
4	Acceptance Preconditions	4
5	Acceptance Procedures	5
5.1	General Requirements	5
5.2	Self-Inspection	5
5.3	Application for Acceptance	5
5.4	Acceptance Preparation	6
5.5	Site Acceptance	6
5.6	Rectification as per Acceptance Comments	8
Appendix A	Archives Acceptance Assessment for Wind Power Projects	9
Appendix B	Main Contents of Self-Inspection Report by the Project Owner	26
Appendix C	Main Contents of Self-Inspection Report by the Supervisor	28
Appendix D	Main Contents of Self-Inspection Report by the Construction Contractor	29
Appendix E	Main Contents of Self-Inspection Report by the EPC Contractor	30
Appendix F	Application Form for Archives Acceptance of Wind Power Projects	31
Appendix G	Main Contents of Archives Acceptance Comments for Wind Power Projects	32
Appendix H	Expert Comments Form for Archives Acceptance of Wind Power Projects	33
Explanation of Wording in This Specification		34
List of Quoted Standards		35

1 General Provisions

1.0.1 This specification is formulated with a view to standardizing the archives acceptance procedure and criteria for wind power projects in accordance with the relevant laws and regulations of China.

1.0.2 This specification is applicable to the archives acceptance for the construction, renovation and extension of wind power projects.

1.0.3 In addition to this specification, the archives acceptance for wind power projects shall comply with other current relevant standards of China.

2 Terms

2.0.1 records of project

all the documents of a project in the form of texts, graphs, audios, and videos generated throughout the project from initiation, approval/record registration, tendering, investigation, design, construction, and supervision to completion acceptance

2.0.2 archival arrangement

process of systematic sequencing of archives including classification, grouping, arranging, numbering and cataloging in accordance with relevant rules

2.0.3 filing of project records

all the arranged documents relating to the project that are transferred to the Project Owner by the Designer, the Construction Contractor and the Supervisor upon the project completion; transfer of all the arranged documents generated at each phase of the project to the archives department by other departments of the Project Owner

2.0.4 archives of project

records of a project that have been appraised, arranged and filed

2.0.5 file

filing unit consisting of a group of related documents

3 Acceptance Organization

3.0.1 Archives acceptance for a wind power project is generally organized by the Project Owner. For a wind power project whose completion acceptance is organized by the State Council or provincial energy administration, the archives acceptance may be organized by the national or local archives administration.

3.0.2 The project archives acceptance team shall be established according to the following requirements:

1. For the project archives acceptance organized by the Project Owner, the acceptance team shall be made up of the persons from the Project Owner and the archives administration as well as technical experts.

2. For the project archives acceptance organized by the state or local archives administration, the acceptance team shall be made up of the persons from the archives administration and technical experts.

3. The number of the project archives acceptance team members should be 5, 7, or 9. The team leader shall be from the archives acceptance organizing authority and the team members shall be made up of archives management experts and wind power experts.

4 Acceptance Preconditions

4.0.1 Wind turbines, substation equipment, power collection lines, central control building, substation building, and road works shall have been constructed, installed, and commissioned according to the design and shall have all been put into operation. The uncompleted works shall not affect the safe and normal operation of the project.

4.0.2 The wind power project shall have passed the 240 h trial running test and have been handed over for production acceptance.

4.0.3 The as-built drawing documents shall have been prepared, and have been reviewed and approved by the Supervisor.

4.0.4 The collection, arrangement, filing and transfer of the records of project have been completed, and the archival arrangement including classification, grouping and cataloging has been basically completed, and shall comply with the current standards of China GB/T 11822, *General Requirements for the File Formation of Scientific and Technological Archives*; DA/T 28, *Requirements for Filing of Records and Archival Arrangement of National Construction of Key Project*; and NB/T 31021, *Specification for Scientific and Technological Records Filing and Arrangement of Wind Farm*.

4.0.5 The Project Owner, the Supervisor, the Construction Contractor, etc. shall have completed their respective self-inspection on the archives of the project and have prepared their self-inspection reports. The Project Owner shall have completed the self-evaluation on archives acceptance, which shall be rated as acceptable.

5 Acceptance Procedures

5.1 General Requirements

5.1.1 The archives acceptance procedure for wind power projects shall include self-inspection, application for acceptance, acceptance preparation, site acceptance, and rectification as per acceptance comments.

5.1.2 Assessment in archives acceptance for wind power projects shall comply with Appendix A of this specification. Archives of a project shall be rated as "acceptable" or "unacceptable". The archives are deemed acceptable if all the dominant items are compliant and the compliance rate of total items reaches 80 % and above.

5.2 Self-Inspection

5.2.1 The Project Owner shall organize the Supervisor, the Construction Contractor, etc. to carry out self-inspection on the archives of the wind power project and prepare their respective self-inspection report. The report should be prepared in compliance with Appendixes B, C and D of this specification. For an EPC wind power project, the self-inspection report shall be prepared by the prime contractor and should comply with Appendix E of this specification. Each party shall be accountable for its own self-inspection report.

5.2.2 The Project Owner shall perform self-evaluation in accordance with Appendix A of this specification.

5.2.3 The Project Owner shall provide supporting documents for archives acceptance, including:

1. Regulations on management of project archives.
2. Records of training and guidance on project archival work.
3. Classification of project archives.
4. Catalog of files and contents of each file.
5. Project breakdown table, list of bidding documents and bids, list of contracts, and list of equipment.
6. Archive-based research and archives accessibility.

5.3 Application for Acceptance

5.3.1 After the project is confirmed satisfying the preconditions for archives acceptance, the Project Owner shall make an application to the archives acceptance organizing authority, which shall be enclosed with an application form and a self-inspection report.

5.3.2 The application form shall be filled out in accordance with Appendix F of this specification.

5.4 Acceptance Preparation

5.4.1 After receiving the application for archives acceptance, the acceptance organizing authority shall conduct a pre-assessment.

5.4.2 In the case of passing the pre-assessment, the acceptance organizing authority shall set up an acceptance team as per Article 3.0.2 of this specification. The acceptance organizing authority shall coordinate with the Project Owner to determine the schedule for site acceptance and issue a notice on archives acceptance for the wind power project.

5.4.3 In the case of not passing the pre-assessment, the acceptance organizing authority shall make comments on rectification and notify the Project Owner.

5.4.4 The Project Owner shall organize the Supervisor, the Construction Contractor, etc. to get prepared for the archives acceptance.

5.5 Site Acceptance

5.5.1 The site archives acceptance for a wind power project shall include the preparatory meeting of the acceptance team, kick-off meeting, site visit, archives inspection, internal meeting of the acceptance team, wrap-up meeting, etc.

5.5.2 Before the kick-off meeting, the acceptance team shall hold the preparatory meeting, which shall be presided over by the team leader and attended by all team members. The preparatory meeting of the acceptance team shall be focused on the acceptance requirements, schedule and task assignment.

5.5.3 The kick-off meeting shall be presided over by the acceptance team leader and attended by all acceptance team members, the Project Owner, the Supervisor, the Construction Contractor, etc. The meeting should:

1. Announce the members of the acceptance team.
2. State the main basis, procedure and work plan for the acceptance.
3. Listen to the Project Owner, the Supervisor, the Construction Contractor, etc. to brief on the project archives management and their self-inspection results.
4. Inquire the concerned parties about their briefings and self-inspection reports.

5.5.4 After the kick-off meeting, the acceptance team shall pay a site visit, understand the project construction and operation conditions, and reconfirm whether or not the project satisfies the acceptance preconditions. If not, the acceptance team shall inform the Project Owner of the conclusion, and reschedule the acceptance once the preconditions are satisfied.

5.5.5 The archive inspection shall include the following:

1. The archives support system and its implementation.

2. The integrity, accuracy and completeness of project archives, as well as the archives transfer and filing formalities.

3. The security, accessibility and informatization of project archives.

5.5.6 The archives acceptance team should check the project archives by means of inquiry, site inspection and spot check. The spot check shall be focused on the project approval documents, design documents, bidding documents and bids, contracts and agreements, documents for concealed works, quality inspection documents, defect remediation documents, supervision documents, as-built drawing documents, equipment documents, etc. The number of files subject to spot check shall not be less than 100.

5.5.7 After the site inspection, the acceptance team leader shall preside over an internal meeting to summarize the inspection results and make an assessment according to Appendix A of this specification. The project archives are deemed acceptable if all the dominant items are compliant and the compliance rate of total items reaches 80 % and above.

5.5.8 The acceptance team shall prepare the acceptance comments based on the assessment, which shall comply with Appendix G of this specification.

5.5.9 The acceptance team members shall fill out the expert comments form as per Appendix H of this specification.

5.5.10 The wrap-up meeting shall be presided over by the acceptance team leader and attended by all acceptance team members, the Project Owner, the Supervisor, the Construction Contractor, etc. The meeting should be held in the following steps:

1. Introduce the implementation of archives acceptance.

2. Read out the acceptance comments.

3. Comment on the identified problems.

4. Solicit the inputs from the parties involved and finalize the acceptance comments.

5　Sign the signature form by all acceptance team members.

6　The Project Owner proposes a rectification plan with respect to the identified problems.

5.6　Rectification as per Acceptance Comments

5.6.1　After the acceptance passes, the acceptance organizing authority shall issue the archives acceptance comments to the Project Owner, and the Project Owner shall rectify the problems accordingly and submit the result of rectification to the acceptance organizing authority for record.

5.6.2　For the project that fails the archives acceptance, the project archives acceptance team shall give the rectification requirements and the Project Owner shall rectify the identified problems within the specified time for review or reacceptance by the acceptance organizing authority.

5.6.3　The acceptance organizing authority shall file the documents and records generated during the process of acceptance.

Appendix A Archives Acceptance Assessment for Wind Power Projects

Table A Archives Acceptance Assessment for Wind Power Projects

S/N	Acceptance item	Acceptance checklist	References	Evaluation criteria	Dominant item	Self-evaluation	Acceptance comments
1		Archives support system and its implementation					
1.1	Organization	(1) The person in charge of the archival work is designated	Related documents	Rated as "unacceptable" if the requirements are not met	—		
		(2) The archival organization or department is set up and staffed with full- and part-time project archivists	Documents about organization setup, personnel responsibilities and training certificate	Rated as "unacceptable" if no archival organization is set up or no full- or part-time archivists are assigned	√		
		(3) A project archives management network including parties involved is established and managed by the Project Owner and the responsible persons are designated	Network charts and documents about implementing the accountability system	Rated as "unacceptable" if the requirements are not met	—		

Table A *(continued)*

S/N	Acceptance item	Acceptance checklist	References	Evaluation criteria	Dominant item	Self-evaluation	Acceptance comments
1.2	Institution	(1) Rules and regulations for project archives management are established, the roles and responsibilities of each department are defined and corresponding control measures are taken	Relevant regulations and rules	Rated as "unacceptable" if necessary regulations are absent	—		
		(2) The table of filing scope and retention period for project records is developed	Table of filing scope and retention period	Rated as "unacceptable" if necessary regulations are absent	—		
		(3) The archive classification scheme and detailed rules for archival arrangement are developed	Related documents	Rated as "unacceptable" if necessary regulations are absent	—		
		(4) The regulations on archive receiving, storing, use, confidentiality, security and statistics are developed	Relevant regulations and rules	Rated as "unacceptable" if necessary regulations are absent	—		

Table A (continued)

S/N	Acceptance item	Acceptance checklist	References	Evaluation criteria	Dominant item	Self-evaluation	Acceptance comments
1.3	Fund	The fund needed for various archival activities of the Project Owner can meet the demand for archiving	Related supporting documents	Rated as "unacceptable" in the case of no archiving fund	—		
1.4	Equipment and facilities	(1) A dedicated archive repository that meets the safekeeping requirement is set up	Field inspection	Rated as "unacceptable" if there is no archive repository	—		
		(2) The equipment and facilities as well as the archives containers in offices and repository meet the requirements of archiving and security	Field inspection	Rated as "unacceptable" if there are quite a lot of problems with the equipment and facilities in offices and repository	—		
1.5	Implementation of management systems	(1) The collection, arrangement, filing and transfer of project records are included in contract management	Related contract terms	Rated as "unacceptable" if not included in the contract management	—		

Table A *(continued)*

S/N	Acceptance item	Acceptance checklist	References	Evaluation criteria	Dominant item	Self-evaluation	Acceptance comments
1.5	Implementation of management systems	(2) The guidance, training and oversight by the Project Owner over the Supervisor, the Construction Contractor, etc. concerning document collection and processing	Related supporting materials	Rated as "unacceptable" if no guidance or training is provided	—		
		(3) The guidance and oversight by the Project Owner over the departments concerned on document collection, processing and filing	Related supporting materials	Rated as "unacceptable" if no oversight or guidance is provided	—		
		(4) The project archival work is kept in pace with project construction and included in the quality management procedure; when project progress and quality are inspected, the collection and processing of project documents are checked; during the completion acceptance of works, section of works and unit of works, the related project documents are inspected or accepted accordingly	Related regulations and records	Rated as "unacceptable" if the project archival work and project construction are out of steps	—		

NB/T 31118-2017

Table A *(continued)*

S/N	Acceptance item	Acceptance checklist	References	Evaluation criteria	Dominant item	Self-evaluation	Acceptance comments
2		Integrity, accuracy and completeness of project archives and transfer and filing formalities					
2.1		Completeness of project archives					
2.1.1	Completeness of power generation documents	(1) Production preparation documents	Related supporting materials	Rated as "unacceptable" if incompleteness exists in the items and the absence rate is 10 % or above	—		
		(2) Production and operation documents at trial running stage			—		
		(3) Production technology documents			—		
		(4) Goods and materials management documents			—		
2.1.2	Completeness of scientific and technological development documents	Scientific and technological development documents	Terms of reference and reports	Rated as "unacceptable" if incompleteness exists and the absence rate is 10 % or above	—		

13

Table A (continued)

S/N	Acceptance item	Acceptance checklist	References	Evaluation criteria	Dominant item	Self-evaluation	Acceptance comments
2.1.3	Completeness of project development documents	(1) Preliminary design documents and project initiation documents relating to land requisition, investment and approval documents	Related documents	Rated as "unacceptable" if incompleteness exists in the items and the absence rate is 10 % or above	√		
		(2) Design documents relating to tender design, construction drawing design and design modification			—		
		(3) Project preparation documents relating to construction land use, tendering, contract and construction commencement report			√		
		(4) Project management documents relating to cost, construction, quality, security, goods and materials, and quality supervision			√		

Table A *(continued)*

S/N	Acceptance item	Acceptance checklist	References	Evaluation criteria	Dominant item	Self-evaluation	Acceptance comments
2.1.3	Completeness of project development documents	(5) Construction documents					
		1) Construction preparation		Rated as "unacceptable" if incompleteness exists in the items and the absence rate is 10 % or above	–		
		2) Foundation construction for wind turbines	Related documents		√		
		3) Construction of central control building and substation			–		
		4) Construction of access works, etc.			–		
		(6) Installation documents					
		1) Installation preparation		Rated as "unacceptable" if incompleteness exists in the items and the absence rate is 10 % or above	–		
		2) Installation of wind turbines and towers	Related documents		√		
		3) Installation of step-up and distribution equipment for wind turbines			–		

Table A *(continued)*

S/N	Acceptance item	Acceptance checklist	References	Evaluation criteria	Dominant item	Self-evaluation	Acceptance comments
2.1.3	Completeness of project development documents	4) Installation of substation equipment	Related documents	Rated as "unacceptable" if incompleteness exists in the items and the absence rate is 10 % or above	√		
		5) Installation of power collection lines			√		
		(7) Supervision documents relating to supervision plan, supervision details, supervision logs, monthly reports, minutes of meetings, control and coordination, etc.			—		
		(8) Documents relating to commissioning, testing and trial running			√		
		(9) Documents such as as-built drawings and the general description			√		
		(10) Completion acceptance documents			—		
		(11) Settlement (final settlement) documents and audit documents			—		
		(12) Audios and videos			—		
		(13) Electronic documents			—		

Table A *(continued)*

S/N	Acceptance item	Acceptance checklist	References	Evaluation criteria	Dominant item	Self-evaluation	Acceptance comments
2.1.4	Completeness of equipment and instrument documents	(1) Documents about wind turbines	Related documents	Rated as "unacceptable" if incompleteness exists in the items and the absence rate is 10 % or above	√		
		(2) Documents about step-up and distribution equipment for wind turbines			—		
		(3) Documents about substation equipment			√		
		(4) Documents about other equipment (water supply and drainage, heating and ventilation, fire protection, special equipment, etc.)			—		
2.2	Accuracy of project archives	(1) The documents are complete and the data are authentic	Related documents	Rated as "unacceptable" if 20 places or more are found inaccurate in each item	—		
		(2) Each type of acceptance assessment table meets the requirements			—		
		(3) The different documents reflecting the same issue are consistent			—		
		(4) The catalog and the hardcopies are consistent and the filed documents are original			—		

Table A *(continued)*

S/N	Acceptance item	Acceptance checklist	References	Evaluation criteria	Dominant item	Self-evaluation	Acceptance comments
2.2	Accuracy of project archives	(5) The as-built drawings are well prepared, reflecting the actual progress of project construction clearly and accurately. The as-built drawing stamping and signing are complete and the Supervisor has reviewed and approved the drawings as required	As-built drawings	Rated as "unacceptable" if 10 places or more fail to reflect the actual situation of project construction, or if the Supervisor fails to perform the review and approval as required	√		
		(6) The documents are legible, the graphs are neat, the review and signing are complete, and the handwriting meets the requirements	Records in the file	Rated as "unacceptable" if 20 places or more fail to meet the requirements	—		
		(7) The description of non-paper documents meets the requirements	Physical archive arrangement	Rated as "unacceptable" if 20 places or more fail to meet the requirements	—		

Table A *(continued)*

S/N	Acceptance item	Acceptance checklist	References	Evaluation criteria	Dominant item	Self-evaluation	Acceptance comments
2.2	Accuracy of project archives	(8) The title of file is concise and accurate, the cataloging is up to the standard and the description is complete and detailed	File title and catalog	Rated as "unacceptable" if there is no file catalog or if 20 places or more have problems in cataloging	—		
		(9) The contents and description of the file are clear and accurate; the page number is accurate and standardized	Contents of file	Rated as "unacceptable" if 10 files or more have no contents or if 20 places or more have problems with the contents of a file	—		
		(10) The file note is filled in and signed as required and the remarks in the file are clear	File note	Rated as "unacceptable" if 10 files or more have no file notes or if 20 places or more have problem with the file note	—		

Table A *(continued)*

S/N	Acceptance item	Acceptance checklist	References	Evaluation criteria	Dominant item	Self-evaluation	Acceptance comments
2.2	Accuracy of project archives	(11) The drawings meet the folding requirements. For the files failing to meet the requirement, necessary remedy measures are taken	Files	Rated as "unacceptable" if 20 places or more fail to meet the requirements	—		
		(12) Files are bound in a secure, tidy and pleasing way. No character is overlapped by any gutter. For individual documents to be filed, the first page of each document shall be stamped with file number. In case that a file consists of drawings only, they need not be bound but shall be stamped with file number one by one	Files	Rated as "unacceptable" if 20 places or more have problem with file binding	—		
2.3	Completeness of project archives	(1) Reasonable classification. Project archive classification plan is developed to ensure that the archives are accurately classified, each type of documents is logically arranged and different types of documents are clearly classified	The classification plan and the classified archives	Rated as "unacceptable" if the archive classification plan is absent	√		

Table A *(continued)*

S/N	Acceptance item	Acceptance checklist	References	Evaluation criteria	Dominant item	Self-evaluation	Acceptance comments
2.3	Completeness of project archives	(2) Reasonable file grouping. It follows the principle of subject-oriented grouping and maintains the logical relationship between documents. The file is of the same subject and has appropriate thickness for easy maintenance and use. For the documents of design change, they shall be arranged in a separate file or files by works, section of works or speciality. The catalog and contents for the file are prepared	File grouping	Rated as "unacceptable" if different categories of documents are arranged in one file and 10 places or more fail to meet the requirements	—		
		(3) Orderly sequencing. Documents with similar content or closely related are sequenced according to their importance or in chronological order; the files under the same subject or a special subject are put together	The sequencing of files and the documents in the files	Rated as "unacceptable" if the files are arranged in disorder or if 10 places or above fail to meet the sequencing requirements	—		
2.4	Filing and transfer	(1) The departments and technical personnel of the Project Owner shall arrange and file the documents under their respective scope according to the requirement	Catalog of files of different types	Rated as "unacceptable" if the absence rate is 10 % or above	—		

Table A *(continued)*

S/N	Acceptance item	Acceptance checklist	References	Evaluation criteria	Dominant item	Self-evaluation	Acceptance comments
2.4	Filing and transfer	(2) The files transferred by the Construction Contractor are reviewed by the Supervisor and go through the formalities as required	Review records	Rated as "unacceptable" if not reviewed	—		
		(3) All the parties involved have transferred the related project records to the Project Owner by works or section of works and completed the transfer procedure	List of transferred files	Rated as "unacceptable" if related project records are not transferred	—		
3	Archive security, accessibility and informatization						
3.1	Archive security	(1) The archive cabinets and shelves are clearly labeled or marked, orderly arranged and reasonably spaced. The archives (repository) keeps a clear record of the archive types and quantity	Repository, register of archives, accession sheet and archival statistics	Rated as "unacceptable" if archive cabinets or shelves are not labeled or marked, or if 10 places or above have problem with cabinet or shelf arrangement, labeling and marking or archive statistics	—		

Table A *(continued)*

S/N	Acceptance item	Acceptance checklist	References	Evaluation criteria	Dominant item	Self-evaluation	Acceptance comments
3.1	Archive security	(2) The archive repository is established, and the regulations on archives management, confidentiality and security are developed, the archive custody is inspected periodically, and measures against fire, theft, direct sunlight, water, humidity, insect, dust and high temperature are taken to ensure archive security	Work log and repository management records	Rated as "unacceptable" if no pertinent regulations are developed or if major hidden danger exists in repository's safety management	√		
3.2	Archive accessibility	(1) The classification catalog and file catalog are complete and standard	Classification catalog and file catalog	Rated as "unacceptable" if classification catalog and file catalog are absent	—		
		(2) Archivists are familiar with the archives and able to locate the archives quickly	On-site witness	Rated as "unacceptable" if slow locating is observed for 10 times or above	—		

Table A (continued)

S/N	Acceptance item	Acceptance checklist	References	Evaluation criteria	Dominant item	Self-evaluation	Acceptance comments
3.2	Archive accessibility	(3) Archive compiling and study is conducted actively	Result of archive compiling and study	Rated as "unacceptable" if the result of archive compiling or study is absent	—		
		(4) The archives statistics is conducted and the annual report is prepared and submitted	Statistics and annual report	Rated as "unacceptable" if statistics or annual report is absent	—		
3.3	Archive informatization	(1) Archive informatization has been in pace with the informatization work of the Project Owner	The performance of archive informatization	Rated as "unacceptable" if archive informatization is not preformed	—		
		(2) Archives management software is provided. File-level and document-level catalog database is established. Full-text digitalization of archives has been performed and plays an important role in archive statistics and access	The use of the software and the database	Rated as "unacceptable" if the archives management software is not provided	—		

Table A (continued)

S/N	Acceptance item	Acceptance checklist	References	Evaluation criteria	Dominant item	Self-evaluation	Acceptance comments
3.3	Archive informatization	(3) Connection with the LAN of the Project Owner is realized and the network service is available. Database security measures are taken	Online operation	Rated as "unacceptable" if no network service is provided	—		

Acceptance criteria: All the dominant items are acceptable, and 80 % or above of the total items are acceptable.

Result of self-evaluation: Among the total 15 dominant items, _____ items are unacceptable; among the total 70 items, _____ items are unacceptable; and the acceptable rate is _____.

Conclusion of self-evaluation: _____

The Project Owner (seal): Date:

Comment on acceptance: Among the total 15 dominant items, _____ items are unacceptable; among the total 70 items, _____ items are unacceptable, and the acceptable rate is _____.

Conclusion on acceptance: _____

Leader of the acceptance team: Date:

NOTES:

1 The acceptance checklist in Item 2.1 is presented according to the categories in Appendix A "Table of Scope of Filing, Archive Classification and Period of Retention for Scientific and Technological Records of Wind Farm" in NB/T 31021, *Specification for Scientific and Technological Records Filing and Arrangement of Wind Farm*.

2 "√" denotes "dominant item".

Appendix B Main Contents of Self-Inspection Report by the Project Owner

1 Project Overview

Project name, project location, project scale, project cost, and parties involved; project milestones, including project approval, construction commencement, grid connection and special items acceptance, and their basic process; project work breakdown and quality acceptance.

2 Project Archives Management

The basis for archives management; the organizational structure of the site archives management office of the Project Owner and the staffing of full- and part-time archivists; the establishment of the archives management system by the Project Owner; the requirements of the Project Owner on the archivist training and filing management for the parties involved; and the control measures taken to ensure the integrity, accuracy and completeness of the filed documents.

3 Project Records Collection, Arrangement and Filing

Scope of filing and classification, requirement on file grouping and file cataloging, preparation and quality of as-built drawings, and the quantity of various types of project archives.

4 Archive Storage and Accessibility

The storage conditions of the archive repository, hardware facilities, safekeeping measures, and the archives accessibility and its effects.

5 Archive Informatization

The archives management software, the full-text digitalization, whether the project archives informatization keeps pace with the informatization work of the Project Owner's company, and whether network service is available.

6 Result of Archive Self-Inspection, Existing Problems and Remedy Measures

Self-inspection on archives conducted by the Supervisor and the Construction Contractor and organized by the Project Owner; the result of self-evaluation on archives as per Appendix A of this specification; problems identified in the archive self-inspection, remedy measures and their implementation.

7 Comprehensive Evaluation

The Project Owner shall evaluate the project archives management, the integrity, accuracy and completeness of the archives, and the security, accessibility and informatization of the archives.

Attachment: Archives Acceptance Assessment for the Wind Power Project

Appendix C Main Contents of Self-Inspection Report by the Supervisor

1 Project Overview

Project location, project scale, scope of work in the supervision contract, etc.

2 Project Archives Management

The basis for archives management, the on-site archives management organization, management staff, archive management system, etc.

3 Supervision Records Collection and Arrangement

The scope of supervision records collection; the classification, grouping and cataloging of the supervision records according to the requirements of the Project Owner; the quantity of the files; and the statistics of files by classification.

4 Technical Review on Project Records

The supervision and guidance provided over the collection and arrangement of the construction documents; the review of the integrity, accuracy and completeness of project documents; the technical review of as-built drawing documents; and the technical review before the transfer of the construction documents.

5 Supervision Records Filing and Transfer

The filing of supervision records and the transfer of the files to the Project Owner, and the quantity of files.

6 Archive Self-Inspection, Existing Problems and Remedy Measures

The self-inspection on files, the problems identified in self-inspection, and the recommendations for remediation.

7 Comprehensive Evaluation

The comprehensive evaluation on the project archives made by the Supervisor.

Appendix D Main Contents of Self-Inspection Report by the Construction Contractor

1 Project Overview

Project location, project scale, scope of work in the contract, etc.

2 Project Archives Management

The basis for archives management, the on-site archives management organization, management staff, archive management system, etc.

3 Construction Document Collection and Arrangement

The scope of construction document collection; the resources provided for document collection and arrangement; the classification, grouping and cataloging of the construction documents; the preparation of the as-built drawings; the quantity of various types of archives and the quantity of files by works or section of works.

4 As-built Drawing Preparation

The general situation of as-built drawing preparation, the review of the as-built drawings, and the quantity of as-built drawings.

5 Technical Review on Construction Documents

The technical review of the construction documents before transfer.

6 Construction Document Filing and Transfer

The filing of construction documents and transfer of the files to the Project Owner, and the quantity of files.

7 Archive Self-Inspection, Existing Problems and Remedy Measures

The implementation of the self-inspection on archives, the problems identified in self-inspection, remedy measures and their implementation.

8 Comprehensive Evaluation

The Construction Contractor's comprehensive evaluation on the project archives.

Appendix E   Main Contents of Self-Inspection Report by the EPC Contractor

1   Project Overview

Project location, project scale, scope of EPC, project management mode, subcontracting, project work breakdown, etc.

2   Project Design and Its Changes

Survey and design reports submitted by the Designer, the compilation of design drawings (atlas), the collection of notices of design change, important design optimization, main design changes during construction, etc.

3   Project Archives Management

Project archives management system, archival organization setup, staffing of full- and part-time archivists, archives management rules, etc.; training of subcontractor archivists on filing management; and measures for ensuring the integrity, accuracy, completeness and security of the project archives.

4   EPC Document Collection and Arrangement

The scope of EPC document collection; the resources provided for document collection and arrangement; the classification, grouping and cataloging of the EPC documents; the preparation of the as-built drawings; and the quantity of files.

5   EPC Document Filing and Transfer

The filing of the EPC documents and transfer of the files to the Project Owner, and the quantity of files.

6   Archive Self-Inspection, Existing Problems and Remedy Measures

The implementation of the self-inspection on archives, the problems identified in self-inspection, remedy measures and their implementation.

7   Comprehensive Evaluation

The evaluation on the integrity, accuracy and completeness of the project archives made by the EPC contractor.

Attachment 1: Archive Statistics

The archive statistics for works and contracts of the project by the type of document and the type of media

Attachment 2: Archives Acceptance Assessment for the Wind Power Project

NB/T 31118-2017

Appendix F Application Form for Archives Acceptance of Wind Power Projects

Form F Application for Archives Acceptance of Wind Power Project

Project name		Project location	
Project Owner			
Approved by		Date of approval	
Total project cost (CNY)		Installed capacity (MW)	
Unit capacity (kW)		Number of wind turbines	
Date of construction commencement		Date of commissioning	
Designed by		Supervised by	
Prime Construction Contractor or EPC contractor			
Wind turbine manufacturer			
Quantity of project files			
Contact person		Tel.	
Address/ZIP code		E-mail	
Description of self-inspection by the applicant		(Company seal) Date:	
Remarks			

31

Appendix G Main Contents of Archives Acceptance Comments for Wind Power Projects

1 Preface

Give a brief description about the basis for acceptance, the acceptance organizing authority and the parties involved, and the acceptance meetings.

2 Project Overview

Project name, project location, project scale, project cost, and parties involved; project milestones, including project approval, construction commencement, grid connection, and special item acceptance, and their basic process; project work breakdown and quality acceptance.

3 Basis and Scope of Acceptance

The laws, regulations, specifications and related approval documents used as the basis for acceptance; and the scope of archives acceptance for the wind power project.

4 Project Archives Management

The archives management organization setup; archives management rules; and measures for ensuring the integrity, accuracy and completeness of the filed documents.

5 Comprehensive Assessment and Conclusion on Project Archives Acceptance

Make a comprehensive assessment on: the creation, collection, arrangement, transfer and filing of project records; the preparation and review of the as-built drawings; the classification and grouping of the archives; and the integrity, accuracy, completeness and security of the archives. Present the conclusion on archives acceptance.

6 Problems and Suggestions for Rectification

Point out the identified problems and make suggestions for rectifications.

Attachment 1 Signature form for archives acceptance team members

Attachment 2 Signature form for representatives of the parties involved

NB/T 31118-2017

Appendix H Expert Comments Form for Archives Acceptance of Wind Power Projects

Form H Expert Comments on Archives Acceptance for Wind Power Projects

Project name		
Description of acceptance		
Overall assessment		
Problems identified		
Suggestions		
Conclusion		
Expert's signature		Date:

Explanation of Wording in This Specification

1. Words used for different degrees of strictness are explained as follows in order to mark the differences in executing the requirements in this specification.

    1) Words denoting a very strict or mandatory requirement:

        "Must" is used for affirmation; "must not" for negation.

    2) Words denoting a strict requirement under normal conditions:

        "Shall" is used for affirmation; "shall not" for negation.

    3) Words denoting a permission of a slight choice or an indication of the most suitable choice when conditions permit:

        "Should" is used for affirmation; "should not" for negation.

    4) "May" is used to express the option available, sometimes with the conditional permit.

2. "Shall meet the requirements of…" or "shall comply with…" is used in this specification to indicate that it is necessary to comply with the requirements stipulated in other relative standards and codes.

List of Quoted Standards

GB/T 11822, *General Requirements for the File Formation of Scientific and Technological Archives*

NB/T 31021, *Specification for Scientific and Technological Records Filing and Arrangement of Wind Farm*

DA/T 28, *Requirements for Filing of Records and Archival Arrangement of National Construction of Key Project*